Actually
Algebra

by Phil Cohen
Copyright 2014 Phil Cohen, all rights reserved.
ISBN: 1494873028
ISBN-13: 978-1494873028

Contents

Introduction 5
Winning money: variables 7
Variable names 9
What we can do with variables 11
Substitution 16
Balance 20
Cancelling 22
Brackets 26
Division 28
More balance 30
Commutation 32
Commutation caution 36
Revision 40
Writing equations from real problems 42
Expansion 46
How do you know what to do? 50
Does algebra always provide a solution? 54
Time to fly solo 56
Negativity 58
Exponents 60

Multiplying and dividing with exponents 66

Algebra with exponents 68

Roots 72

Fractions 74

Multiplication signs 78

Algebra with fractions 80

Final example 82

What's next? 84

Introduction

I'm going to make some assumptions in writing this book ... if I've got any of them wrong, perhaps this isn't the book for you. I assume you:

- hated mathematics by the time you left school (even if you loved it at some point before that)
- need to learn some algebra for your job, or for study
- don't have a lot of time
- aren't really interested in the history of algebra, or stories of famous mathematicians, and aren't ready to be convinced that algebra is fun
- have a sense of humour and don't mind me spelling it that way

If you're still with me, let's get started.

Winning money: variables

Remember those quiz games where there's $10,000 to be given away, and the contestant has to choose a briefcase, and depending on which one they choose they get a different amount of money? If you don't remember the show that's okay - just imagine it.

Each briefcase is labelled either "A", "B" or "C". If you choose briefcase "A" you know it will contain some number of dollars ... but you don't know how many. Same with B or C.

That's what a **variable** is. It's a briefcase with a label on it, and inside that briefcase is a number. And you don't know what that number is, yet.

Now, the number won't change, so it's actually misleading to call it a 'variable', because it's not variable - it's fixed. It's actually better to call it an 'unknown', because that's what it is. It's a number we just don't know yet.

So what's the point of labelling a number that we don't know?

Well, even though we don't know what the number is, we might know something about it. For example, in the quiz we know the total will be 10,000 (dollars, right?). There might be other things we know about it, but we'll come to that later. For the time being, just remember that a 'variable' has a label (like "A") and has **an actual value that we don't yet know**.

Another example: the temperature outside today is a number, but I don't know what that number is. Let's call it "T". So we don't know what the number "T" is, but we have a label for it. "T" is like another briefcase with an unknown number in it - but instead of a number of dollars it's a number of degrees.

Or I might measure your heart rate. We'll call this "H". Again, it's just a number with a label:

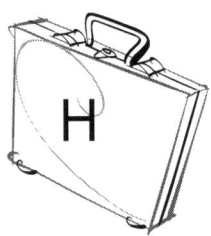

Variable names

Usually people choose labels that stand for something - T for Temperature, H for Heart rate - because they're easy to remember, easy to work with. But not always.

As in any other place where humans work, there are *traditions* in variable names. In electronics the variable name for Voltage is "V", and for Resistance is "R" - logical enough. But the variable name for Current is ... "I"! Why? No good reason, just tradition.

Some people (particularly math teachers, for some reason) like to use "x" and "y" a lot. No particular reason. They're just labels.

Mathematicians, when they're trying to look clever, or when they've run out of English letters as variable names, will use letters of the Greek alphabet. That's exactly equivalent to taking your shoes and socks off to count to more than 10.

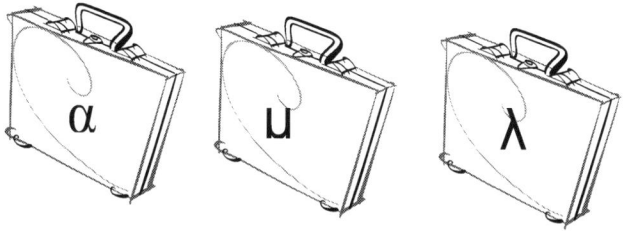

My favourite kind of variable name is a complete word - like "Temperature", or "Heart rate". There's nothing mathematically incorrect about that, and it makes it really easy to work with. A bit more typing, but I can live with that.

What we can do with variables

So now we know that a variable is just a label for a number we don't yet know:

What use is that? Well, sometimes we know something about these variables even though we don't know their value. For example, remember the quiz show? There was a total of $10,000 in cash to be won, and it was split in some way between three briefcases:

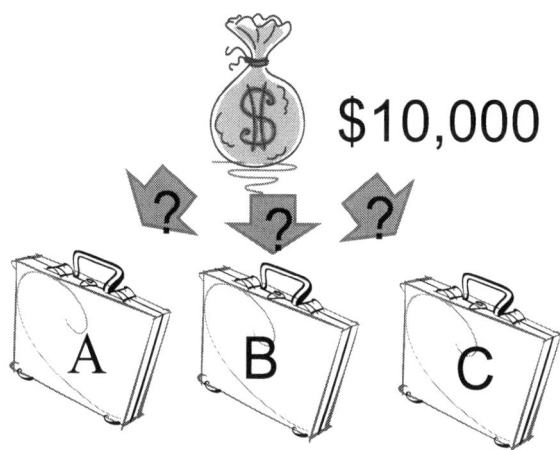

We know that the total amount in all three briefcases is 10,000 (dollars). So if we take the value of A (ie the amount in briefcase A) and the value of B and the value of C and add them together, we'll get 10,000.

If we add together A + B + C we know that's going to be 10,000 total. In English we might say "A plus B plus C equals 10,000". Or as an **equation** it looks like:

A + B + C = 10000

An equation is just a shorthand way of writing down a sentence that contains the word 'equals'.

So we've got three variables (A, B and C) that we don't yet know the value of (we don't know how much is in briefcase A, do we?) but know *something* about them: there's $10,000 in there in total.

Now the point of algebra is to find the *actual* values of the variables.

So although we don't know how much is in each briefcase yet, we might be able to find out using algebra. And that's what it's all about: you use equations to write down what you do know about some numbers, when you don't yet know their values. Then you use algebra to find the actual values. Then you're done.

Back at our quiz game. We still don't know what's in each briefcase yet. But what if we were told that there was three times as much in B as there was in A, and twice as much in C as there was in B. If you're good with puzzles you can work out that there's got to be $1000 in A. But there's an easier way to work it out without doing your head in: with algebra.

Let's write down what we know:

A + B + C = 10000

B = 3xA ("three times as much in B as in A")

C = 2xB ("twice as much in C as in B")

If there's three times as much in B as there is in A, and twice that in C, then there must be *six* times as much in C as there is in A:

B = 3xA

C = 2xB

so

C = 6xA

(If your eyes just glazed over, take the time to look at this again carefully until you understand it, because what you've just done is to use a *belief* that you learned at school "I don't understand algebra" and applied it in an inappropriate setting. There's nothing complicated about this stuff. If it helps, read it out loud. Remember: you're older now, and you *need* this stuff, otherwise you wouldn't be reading this book. Leave your inappropriate beliefs in the past where they belong.)

So we now know that:

B = 3xA

and

C = 6xA

And we know that:

A + B + C = 10000

Let's replace the 'B' by '3xA'. We know we can do this because we know that B = 3xA (in other words, whatever number B is, it's three times the number that A is).

So instead of:

A + **B** + C = 10000

We can write:

A + **3xA** + C = 10000

Now that's the cool thing about algebra. Because we know that A + B + C = 10000 and we know that B = 3xA we can say for **absolute certain** that A + 3xA + C = 10000.

Now we can do the same thing for the C:

A + 3xA + **C** = 10000

and

C = 6xA

so

A + 3xA + **6xA** = 10000

So what does this tell us? Well, A + 3xA + 6xA is just 10xA (just because 1 + 3 + 6 is 10).

So instead of:

A + 3xA + 6xA = 10000

we can just write

10xA = 10000

And if 10 times a number is $10,000, then the number must be $1000.

A = 1000

So we know that briefcase "A" holds $1000.

Exercise

Your turn. Work out how much is in briefcases B and C. Then add the total amount in A, B and C to make sure you've got the right answer (the total should be $10,000).

Answers

Briefcase B has $3000 and briefcase C has $6000.

Explanation

We already know that there's three times as much in B as there is in A, so there must be $3000 in B. And we know that there's twice as much in C as there is in B, so there must be $6000 in C.

Or from the algebra:

B = 3xA

A = 1000

so

B = 3000

And also:

C = 2xB

so

C = 6000

Substitution

There's a fancy name for what we just did to find out A, B and C. We used what's called **substitution**.

When we went from this:

A + **B** + C = 10000

B = 3xA

to this:

A + **3xA** + C = 10000

what we did was to *substitute* for B. It's as if we took "B = 3xA" and zotted it in instead of "B" in the other equation.

<show "B = 3xA" being inserted into "A + B + C = 10000">

We did the same thing when we 'substituted' C = 6xA for C.

Exercise

Here's an equation:

A + B = 6

Here's another equation:

B = 5xA

Substitute the second equation into the first one and find out the value of A. If it helps, think of the variables as briefcases containing money!

Answer

A = 1

Explanation

(But try it yourself first before you read this!)

The first equation is:

A + B = 6

Instead of B we substitute 5xA (because we know that B = 5xA, so B and 5xA must be the same number):

A + **B** = 6

B = 5xA

A + **5xA** = 6

Now A + 5xA is just going to be 6xA (because 1 + 5 is 6), so:

6xA = 6

Well, if six times something is six, then the something must be 1, right? So:

A = 1

And by the way because B = 5xA, we can figure out that B = 5 as well.

Let's check that against the original equations, which are:

A + B = 6

and

B = 5xA

If A is 1 and B is 5 then A + B will indeed be 6 (so the first equation holds). And if B is 5 and A is 1 then the second equation is okay as well.

Exercise

(In this book, the first exercise will be pretty much a repeat of what you've already been shown, but the second exercise will be a little, tiny, bit of a stretch.)

Here are some equations:

A x B = 10

B = 2

Substitute the second equation into the first and find A.

Answer

A = 5

Explanation

(Again, please try it yourself before you read the explanation).

So here's the first equation:

A x **B** = 10

But because we know that **B = 2**, we can substitute 2 instead of B:

A x **2** = 10

So if a number times 2 is 10, then that number must be ... 5.

Exercise

(And a little bit more of a stretch ...)

B + C = 90 - A

A = 10

B = 3xC

What is C?

Answer

C is 20

Explanation

B + C = 90 - **A**

A = 10

so

B + C = 90 - **10**

so

B + C = 80 (because 90 - 10 is 80)

Substituting for B:

B + C = 80

B = 3xC

3xC + C = 80

so

4xC = 80

so

C = 20

Balance

Back to our quiz show ... we're going to introduce a new 'joker' briefcase labelled J. This may have $100,000 in it, or $1. We don't know until we open it.

So now we still have briefcases A, B and C which share the $1000 prize money, plus the J briefcase which may have either $100,000 or $1. How much is there in total?

Well, we know that:

A + B + C = 10000

But we want to find out A + B + C + J (because that's the total cash on offer). All we really know is that this will be the original $10,000 *plus* whatever's in J:

A + B + C + J = 10000 + J

Notice that this will be true no matter what the value of J is. If J has $100,000 it will be true, and if J has $1 it will still be true.

So we started with this equation:

A + B + C = 10000

And we ended up with this:

A + B + C + **J** = 10000 + **J**

It's as if we just added 'J' to both sides of the equation. By 'side' I mean either the bit to the left of the = sign (the 'left hand side' of the equation) or the bit to the right of the = sign (the 'right hand side' of the equation).

In fact, there's a general rule that: **if you add the same thing to both 'sides' of an equation, you won't break it**.

So if we know that:

A + B + C = 10000

then we can add anything we like to both 'sides' and it will still be true:

A + B + C + **J** = 10000 + **J**

A + B + C + **X** = 10000 + **X**

A + B + C + **493** = 10000 + **493**

A + B + C + **Temperature** = 10000 + **Temperature**

No matter what you add (and no matter what value it has), as long as you add it to both sides of the equation all will be well.

The same goes for subtracting things:

A + B + C - J = 10000 - J

A + B + C - X = 10000 - X

A + B + C - 493 = 10000 - 493

A + B + C - Temperature = 10000 - Temperature

Cancelling

Now this idea that you can add or subtract anything from an equation as long as you do the same to both sides is a very useful one.

Look at this:

A + B = 10000 - C

It looks sort of familiar, but C is on the wrong side. How can we get it where we want it (so we can have an equation for A + B + C?). Well, remember that we can add anything we like to both sides.

What if we add an extra C to both sides?

A + B = 10000 - C

A + B + **C** = 10000 - C + **C**

Now we have what we want on the left hand side. What about the right hand side?

Well, 10000 - C + C is going to be just 10000. Whatever C is, if we subtract it and then add it again we'll be back with the value we started with, 10000. So we can 'cancel' the two Cs on the right hand side, like this:

A + B = 10000 - C

A + B + **C** = 10000 - C + **C**

A + B + C = 10000 ~~- C + C~~

A + B + C = 10000

Now we have C back where we want it, alongside A and B.

We can use this to 'move' variables from one side of an equation to the other.

Look at this:

A + B = 10 + D

B = 3xA

D = 2xA

What's A?

Let's start by substituting for B:

A + **B** = 10 + D

B = 3xA

A + **3xA** = 10 + D

So far, so good. Now we'll substitute D as well:

A + 3xA = 10 + **D**

D = 2xA

A + 3xA = 10 + **2xA**

In order to work out what A is, we really have to get that 2xA from the right-hand side over to the left-hand side so we can add it to the other 'A's. So let's subtract 2xA from *both* sides:

A + 3xA = 10 + 2xA

A + 3xA - **2xA** = 10 + 2xA - **2xA**

Now we can cancel on the right-hand side:

A + 3xA - 2xA = 10 +~~2xA - 2xA~~

A + 3xA - 2xA = 10

So now we can see that there are a total of 1 + 3 - 2 = 2 'A's on the left-hand side:

2xA = 10

A = 5

Exercise

B + A = 3 + A

What is B? (I know it's obvious, but use algebra to prove it!)

Answer

B = 3

Explanation

B + A = 3 + A

Subtract A from both sides:

B + A - **A** = 3 + A - **A**

Cancel on the left:

B + ~~A - A~~ = 3 + A - A

B = 3 + A - A

Then cancel on the right:

B = 3 + ~~A - A~~

B = 3

Exercise

B + A = 8

A = B + 2

What is A?

Answer

A = 5

Explanation

B + A = 8

A = B + 2

Substitute for A in the first equation:

B + **A** = 8

A = B + 2

B + **B + 2** = 8

Subtract 2 from both sides:

B + B + 2 **- 2** = 8 **- 2**

Cancel on the left:

B + B ~~+ 2 - 2~~ = 8 - 2

B + B = 8 - 2

Of course, 8 - 2 is just 6, so:

B + B = 6

So B must be 3. And since B + A = 8, A must be 5.

Brackets

Brackets (sometimes called braces for some reason) are your friend.

Why do we need them in algebra? Well take a look at this:

4 + 2 x 3

There are two ways to work this out. First, Add 4 + 2, then multiply by 3:

4 + 2 x 3

6 x 3

18

The other way is to multiply 2 x 3 first, then add 4:

4 + 2 x 3

4 + 6

10

Now, I know there are rules about this (do the multiplication first, and so on) but brackets make the whole thing much clearer. So in this case we could either write:

(4 + 2) x 3

or

4 + (2 x 3)

... depending on what we actually meant. In fact when you're writing equations it's good practice to *always* use brackets for clarity, even when the 'multiply first' rule makes it obvious to you what's going on. It's always obvious until you make a mistake!

So if you write something like this:

A + B x C = D

... make sure you either write it as:

(A + B) x C = D

or

A + (B x C) = D

Division

There are a number of different ways of showing that you want to divide one number by another. You can write:

2 ÷ 3

or you can write:

$$\frac{2}{3}$$

or you can write:

2 / 3

... but they're all just different ways of saying **exactly the same thing**. In this book I'll use the 2 / 3 format.

More balance

So we can add or subtract anything we like from an equation, as long as we do the same on both sides, and we can remove things that cancel each other out. What about multiplying and dividing? We can take this:

A x B = D

and divide both sides by B:

A x B / B = D / B

... then cancel on the left hand side (because sometimes times B, divided by B is just the original something):

A ~~x B / B~~ = D / B

A = D / B

Just the same as we can with adding or subtracting, **we can multiply or divide as long as we do the same to both sides, and we can cancel multiply and divide**.

Let's look at an example:

A x 4 = C x 8

A = 2 + C

Substitute for A in the first equation:

(2 + C) x 4 = C x 8

Notice the brackets ... just to make sure what we mean. Now divide both sides by 4:

(2 + C) x 4 **/ 4** = C x 8 **/ 4**

Cancel the 4s on the left-hand side:

(2 + C) ~~x 4 / 4~~ = C x 8 / 4

(2 + C) = C x 8 / 4

Now just divide 8 by 4 on the right hand side to get 2:

(2 + C) = C x 2

Just ditch the brackets because we don't need them any more:

2 + C = C x 2

Now subtract C from both sides:

2 + C - **C** = (C x 2) - **C**

2 +~~C~~ ~~C~~ = (C x 2) - C

2 = (C x 2) - C

And "(C x 2) - C" just means "two Cs minus one C", which gives us:

2 = C

Commutation

No, it has nothing to do with how you get to work. It's just a fancy way of saying that you can swap things around a bit if you need to.

It means that A + B is the same as B + A ... seems obvious, really. It also means that A x B = B x A. So when you have an equation like this:

A + B + C = 4

You can swap it around all you like without breaking anything:

B + A + C = 4

C + A + B = 4

And when you have something like this:

A x B x C = 9

You can do the same thing:

B x A x C = 9

C x B x A = 9

So we can take this equation:

A + B = B + C + 2

and subtract B from both sides:

A + B - **B** = B + C + 2 - **B**

then cancel the Bs on the left hand side:

A + ~~B - B~~ = B + C + 2 - B

A = B + C + 2 - B

Then move things about a bit on the right hand side, like this:

A = B + C + 2 - B

A = C + 2 + B - B

So that we can now cancel on the right as well:

A = C + 2 ~~+ B - B~~

Caution! When you get something like this:

A + B x 2 = 9

You *can't* swap the A and B around! That's another good reason to always put in the brackets:

A + (B x 2) = 9

... so you can see that you can't just swap the A and B, because they're not really next to each other.

Example

A + B = 8

A = B + 2

What is A?

Answer

A = 5

Explanation

Substitute for A:

A + B = 8

A = B + 2

B + 2 + B = 8

Now subtract 2 from both sides:

B + 2 + B **- 2** = 8 **- 2**

Move things about a bit:

B + B + 2 - 2 = 8 - 2

Cancel the 2s on the left:

B + B +~~2~~ ~~2~~ = 8 - 2

B + B = 8 - 2

Now do the sum on the right:

B + B = 6

So B must be 3, and since:

A + B = 8

... then A must be 5. If you've been doing the examples, you'll notice that this is actually the same as an earlier one, but with some of the letters swapped so that you had to use commutation to sort them out.

Commutation caution

Be careful with things like this:

A - B + C

When you reorder them, think of the '+' and '-' signs as being 'stuck' to the letters that follow them, like this, and add an 'invisible' + sign to the first one, like this:

+A -B +C

... so that when you move them around the signs move with the letters:

+A +C -B

+C +A -B

-B +A +C

Actually there's nothing wrong with writing:

+A - B + C

... but people tend to leave off the '+' if it's at the start of an equation. So if you reordered A - B + C to give:

+C -B +A

then you can write it as:

+ C - B + A

if you like, or just as:

C - B + A

Notice that this doesn't apply to a '-' sign if it's at the front; you have to leave it there:

-B +A +C

needs to be written as:

-B + A + C

The reason for this is that '+' is the 'default' sign that's sort of in front of anything that doesn't have a '-' sign. Sometimes we write that '+' in, sometimes we don't.

The same thing applies to 'x' and '/' signs. If you're reordering this:

A x B / C

Then you'll need to add an 'invisible' x to the first letter, and 'stick' the 'x' and '/' to the letters that follow them:

xA xB /C

So now you can reorder them as you like:

/C xA xB

xB xA /C

If you find a 'x' sign at the front you can just remove it. But if you find a '/' at the front you'll have to do something a bit strange ... you'll have to add a '1' to the start and put brackets round it, so this:

/C xA xB

has to be written:

(1 / C) x A x B

Why? Well, /C means "divided by C" and since there's nothing to the left of it for it to divide, you have to stick in a '1' so that it makes sense as "1 divided by C". And the brackets? Well, they're just there to avoid confusion (believe it or not).

Just to prove to you that this all works, let's look at a simple example:

4 x 6 / 2

The answer to this is just 2 (4 x 6 = 24, and 24 / 2 = 12). What if we changed the order and put the 2 first:

4 x 6 / 2

x4 x6 /2

/2 x4 x6

Which we'll have to write as:

(1 / 2) x 4 x 6

Is this still the same answer as before? Let's check.

1/2 is just a half. A half times 4 gives us 2, and 2 times 6 gives ... 12. The same answer as before; so the method works.

Remember that you only have to do the trick with the 1 and the brackets if it's a '/' that ends up at the start. If it's a 'x' you can just remove it.

Example

Rearrange B - A - C in alphabetical order.

Answer

-A + B - C

Explanation

Starting with:

B - A - C

Add an 'invisible' + to the first letter:

+B - A - C

Then stick the signs to the letters that follow them:

+B -A -C

Now rearrange into alphabetical order:

-A +B -C

And now unstick the signs:

- A + B - C

('Stuck' and 'unstuck' signs are sometimes called 'monadic' and 'dyadic' but only by computer programmers.)

Example

Rearrange B / A x C in alphabetical order.

Answer

(1 / A) x B x C

Explanation

So we start with:

B / A x C

First, put a 'x' at the start and stick the signs onto the letters that follow them:

xB /A xC

Now rearrange into alphabetical order:

/A xB xC

Now unstick the signs, but add a '1' and some brackets to the A:

(1 / A) x B x C

Revision

We've covered a lot of ground so far, so here's a reminder of where we've been:

- Variables are just labels covering numbers that we don't yet know.
- The purpose of algebra is to find those numbers.
- You can substitute one equation into another one.
- You can add, subtract, multiply or divide one side of an equation as long as you do the same to the other side as well.
- If you get two variables that cancel each other out, you can remove them.
- Brackets avoid confusion.
- You can change the order of variables as long as you stick to some rules.

Writing equations from real problems

In order to actually use algebra, you'll often have to take a real-world problem and turn it into a set of equations. The first step in this is to come up with some variable names (pick ones that are easy to use and easy to remember). Then just translate what you know about those variables into equations.

Here's an example.

Joe, Mary and Fred are all siblings. Fred is ten years older than Mary. Mary is twice Joe's age and Joe is one third the age of Fred. How old is Fred?

First let's choose some variable names. There's a special form of words that you use for this:

"Let J be Joe's age.

Let M be Mary's age.

Let F be Fred's age."

Nothing magical about that, just yet another tradition.

Now let's write down what we know as some equations:

"Fred is ten years older than Mary" give us:

F = M + 10

"Mary is twice Joe's age" gives us:

M = 2 x J

"Joe is one third the age of Fred" gives us:

J = F / 3

So we have three equations. I'll let you try to figure out the answer if you like, but at this point all we're trying to do is to write some equations. (Answer: Joe is 10).

Example

Write this down as a set of equations:

"Anthony is three years older than Sian, and Sian is half the age of Joanne. Anthony is 21". How old is Joanne?

Answer

Let A be the age of Anthony

Let S be the age of Sian

Let J be the age of Joanne

$A = S + 3$

$S = J / 2$

$A = 21$

Joanne is 36.

Example

Write this down as a set of equations:

A box weighs 14 kg when it contains a lamp, a chair and a ball. The ball weighs 1 kg, the empty box weighs 1 kg, and the chair is twice the weight of the lamp. How much does the lamp weigh?

Answer

Let BE be the weight of the empty box

Let BF be the weight of the full box

Let L be the weight of the lamp

Let C be the weight of the chair

Let B be the weight of the ball

BF = 14

BF = BE + L + C + B

B = 1

BE = 1

C = 2 x L

The lamp weighs 4 kg.

Explanation

This is a little more tricky. For one thing, you aren't actually told that the weight of the full box is the weight of the empty box plus the stuff in it. Pretty obvious, but not explicitly stated in the problem.

Second, notice how we've chosen variable names for *everything* including the box when it's empty and the box when it's full. You'll find that having too many variable names is better than having too few.

Example

Write this down as a set of equations ("scribble" is not a real game):

In a game of scribble each team gets two points for each line that they win, and three points for each of their opponent teams' penalties. Each game consists of four lines. Glasgow Scribblers won every line, and got a total score of 11 for the game. How many penalties were there?

Answer

Let G be the score for a game

Let L be the number of lines won in a game

Let P be the number of penalties in a game

G = 2xL + 3xP

L = 4

G = 11

There was one penalty.

Expansion

If you have a cup and a saucer in each place setting, and you have four place settings, how many cups do you have? Obviously, the answer is four.

That's a bit like this:

4 x (Cup + Saucer)

Which is just:

4 x Cup + 4 x Saucer

What we've just done is to **expand** the expression by multiplying each of the things inside the brackets individually.

So if we expand:

5 x (A + B + C)

we will get

5xA + 5xB + 5xC

What if we expand this?:

Z x (A + B)

We'll just get:

ZxA + ZxB

Makes sense?

Example

A = C x (B + 3)

A = (B x C) + 3

What is C?

Answer

C = 1

Explanation

A = C x (B + 3)

Let's *expand* this to:

A = C x B + C x 3

Now we'll *substitute* for A from the second equation:

A = C x B + C x 3

A = (B x C) + 3

(B x C) + **3** = C x B + C x 3

Rearrange C x B so it matches what's on the left:

(B x C) + 3 = **C x B** + C x 3

(B x C) + 3 = **B x C** + C x 3

Now rearrange both sides so that the B x C is at the end:

(B x C) + 3 = **B x C** + C x 3

(B x C) + 3 = C x 3 + **B x C**

(B x C) + 3 = C x 3 + B x C

3 + (B x C) = C x 3 + B x C

We can put in some extra brackets on the right just to make the next step a bit clearer:

3 + (B x C) = C x 3 + **(**B x C**)**

Now subtract (B x C) from both sides:

3 + (B x C) - (B x C) = C x 3 + (B x C) - (B x C)

And cancel on the left:

3 + ~~(B x C)~~ ~~(B x C)~~ = C x 3 + (B x C) - (B x C)

3 = C x 3 + (B x C) - (B x C)

Then cancel on the right:

3 = C x 3 + ~~(B x C) – (B x C)~~

3 = C x 3

So if C times 3 is 3, C must be 1.

How do you know what to do?

If you've been following all of these complicated steps (and I hope you have) then by now you'll be asking "How do I know to do *these* steps in *this* order to get the answer?". Well, it's a bit like solving a puzzle: you have to work backwards from where you want to get to.

So let's look at the example we've just seen:

$A = C \times (B + 3)$

$A = (B \times C) + 3$

What is C?

The basic rule is **first expand everything, then try to get rid of all of the letters except the one you're looking for** (in this case, C because the question we're trying to answer is "What is C?").

When I was solving this my thinking went like this:

$A = C \times (B + 3)$

$A = (B \times C) + 3$

Hmm ... okay let's start by expanding $C \times (B + 3)$ in that first equation:

$A = C \times B + C \times 3$

$A = (B \times C) + 3$

Would be nice to get rid of the A somehow - and I can do that by substituting the second equation into the first:

$(B \times C) + 3 = C \times B + C \times 3$

Great, that's A got rid of. Now what about B? Oh look, I can cancel the C x B from both sides and that will get rid of the Bs:

3 = C x 3

Well now I'm left with just C and I can see the answer.

Example

11 + A = 4 x (A + B)

B = 2 x A

Let's start by expanding the first equation:

11 + A = 4 x (A + B)

11 + A = 4 x A + 4 x B

Now let's get rid of the B by substituting:

11 + A = 4 x A + 4 x **B**

B = 2 x A

11 + A = 4 x A + (4 x **2 x A**)

Let's just tidy that up:

11 + A = 4 x A + (4 x 2 x A)

11 + A = 4 x A + (8 x A)

11 + A = 12 x A

Let's get the As all on the same side of the equation:

11 + A = 12 x A

Subtract A from both sides and then cancel:

11 + A = 12 x A

11 + A - A = 12 x A - A

11 + ~~A - A~~ = 12 x A - A

Now tidy up the right hand side:

11 = 11 x A

So A must be 1.

Does algebra always provide a solution?

I'd hate you to get the impression that there's some sort of handle you can turn to solve every set of equations. Often when I'm faced with a problem I'll have to take several attempts, try different things, until I finally get the right answer. The nice thing about algebra, though, is that (as long as you're careful not to make any errors) it won't give you the wrong answer, ever. However, it can take you some time to get there.

Solving a set of equations is a bit like doing a puzzle: you have to try stuff, backtrack, scratch your head and try again. Then screw up that bit of paper and chuck it across the room and take a new one. But when you finally get the answer, it's like completing a puzzle!

For **real-world problems** (as opposed to problems you'll come across in a book), sometimes there is no answer. Some equations just can't be solved: but the people who write math textbooks just don't put them in the book.

Don't believe me? Here's one that you can't solve:

A + B + C + D = 49

C = 4

D = 5

What is A?

Time to fly solo

Okay, I know that some of you will have read through this whole book without actually having tried any of the examples. But here are some problems for you to solve all on your own - I've given just the answer so that you can check that you've got it right, but I won't give you an explanation. Remember that you probably won't get any of these first time through - just keep trying!

Example 1

A = 4 + B

B = 3

What is A?

Example 2

A = B + 4 x B

B = 6

What is A?

Example 3

A = B + C - 3

B = C

C = 2

What is A?

Example 4

A = 4 x (A + B) - 22

B = 2 x A

What is A?

(This last one's a little tricky, but keep trying. Don't give up for at least 30 minutes.)

Example 5

There are 30 guests in a club, and they all paid to get in. The early bird price (before midnight) was $20, and the rest paid $50. When I started performing at the club at exactly midnight there were only 25 people in the place, including the band (4 members) and me. We got half of the take for the night and split it between us. How much did I get? (The band members and I didn't have to pay to get in, and we got free drinks).

Answers

Example 1: A = 7, Example 2: A = 30, Example 3: A = 1, Example 4: A = 2, Example 5: $90 plus the free drinks.

Negativity

Just in case you've forgotten, I'm going to go over negative numbers a bit.

Now, when you look at a ruler you'll see the numbers on it are in sequence:

1 2 3 4 5 6 7 8 9 10

So far, so good. It's pretty obvious what the next number in that sequence is - 11.

But what if you wanted to extend the sequence to the left? Well, you can see that as you move to the left each number gets smaller by 1. So the number to the left of the ruler will be 1 - 1 = 0:

0 1 2 3 4 5 6 7 8 9 10

Kind of makes sense, really. What's the number to the left of the 0?

We know it's going to be 1 less than 0, so we'll have to invent something called "negative one", which is 1 less than 0:

-1 0 1 2 3 4 5 6 7 8 9 10

If you like, you can think of -1 as meaning "take away 1", because that's really what it is. So what would be to the left of -1? It would be "take away 1" take away one, which is just "take away two":

-2 -1 0 1 2 3 4 5 6 7 8 9 10

And in fact we can extend this sequence in both directions as far as we like:

-10 -9 -8 -7 -6 -5 -4 -3 -2 -1 0 1 2 3 4 5 6 7 8 9 10

This idea is called the **number line**, and it's just a way of thinking about negative numbers (they're the ones to the left of the zero).

So how do you use these negative numbers? Well, you've been using them already. Look at this:

4 - 3

That's actually the same as:

4 -3

(Read this as "four and minus three").

Although to be completely correct we'll have to show it like this:

4 + -3

It actually means the same as 4 - 3. So in a sense you know how to use negative numbers (like -3). They just mean "take away this value".

Exponents

Okay, if you've come this far and managed to do all of these examples you're doing well.

Here's something you might have seen before somewhere:

3^2

It means 9 ... the little '2' at the top right means "squared" or "times itself", so the answer is 3 x 3 = 9.

Example

What's 5^2?

Answer

25

Explanation

5 x 5 = 25

So far so good. Now what do you think this means?:

2^3

Well, if 2^2 is 2 x 2, so 2^3 is just 2 x 2 x 2.

And 2^4 is 2 x 2 x 2 x 2.

And 2^5 is 2 x 2 x 2 x 2 x 2, and so on.

Example

What's 3^4?

Answer

81

Explanation

3^4 is 3 x 3 x 3 x 3, which is 81.

Let's look at a sequence of these things and see if we can figure out what happens at both ends.

3^4 is 3 x 3 x 3 x 3 = 81

3^3 is 3 x 3 x 3 = 27

3^2 is 3 x 3 = 9

3^1 is 3 = 3

Huh? That last one is a bit strange, but if you look at the pattern, you'll see it makes sense. Each time you go down the pattern, the number of 3s reduces by one.

What about 3^0? That would have ... zero threes. It will look like this:

3^0 is

But that doesn't look right. Notice also that as you go down the list each answer (81, 27, 9, 3) is a third of the one above? So the entry for 3^0 should actually be 3 / 3, which is just 1:

3^4 is 3 x 3 x 3 x 3 = 81

3^3 is 3 x 3 x 3 = 27

3^2 is 3 x 3 = 9

3^1 is 3 = 3

3^0 is 1

Can we go down any further? Yes, we can. Remember the ruler from a page or so back? As we went lower than 0 we got to negative numbers. The same happens with this, so the next answer down is 3^{-1}.

And what would be the value of that? Well, remember that the values are each a third of the one before: 81, 27, 9, 3, 1. So the next answer must be a third of one, which is 1/3:

3^4 is 3 x 3 x 3 x 3 = 81

3^3 is 3 x 3 x 3 = 27

3^2 is 3 x 3 = 9

3^1 is 3 = 3

3^0 is 1

3^{-1} is 1/3

And the next one will be a third of that:

3^{-2} is 1/3 x 1/3 = 1/9

And so on:

3^4 is 3 x 3 x 3 x 3 = 81

3^3 is 3 x 3 x 3 = 27

3^2 is 3 x 3 = 9

3^1 is 3 = 3

3^0 is 1

3^{-1} is 1/3

3^{-2} is 1/3 x 1/3 = 1/9

3^{-3} is 1/3 x 1/3 x 1/3 = 1/27

3^{-4} is 1/3 x 1/3 x 1/3 x 1/3 = 1/81

Now this is kind of interesting, but the title of this book is "Actually useful", so what can we do with this?

Well, the main application is with 10 rather than 3. The list for 10 looks like this:

10^4 is 10 x 10 x 10 x 10 = 10,000

10^3 is 10 x 10 x 10 = 1000

10^2 is 10 x 10 = 100

10^1 is 10 = 10

10^0 is 1

10^{-1} is 0.1

10^{-2} is 1/10 x 1/10 = 0.01

10^{-3} is 1/10 x 1/10 x 1/10 = 0.001

10^{-4} is 1/10 x 1/10 x 1/10 x 1/10 = 0.0001

Now in science and engineering, you're often dealing with very large or very small numbers. For example, the number of atoms in a mol is called "Avogadro's number" and it's about 600000000000000000000000. Now even if I put the commas in, that's still pretty hard to read: 600,000,000,000,000,000,000,000. But if I write it like this, it's a bit easier:

6×10^{23}

Now, what does 10^{23} mean? If you have a look at the table above, you'll see that the number of zeros in the result is the same as the number: 10^4 is 10,000, which has four zeros, 10^2 is 100 which is two zeros. So 10^{23} is just 100000000000000000000000. Simple, really.

The name for the little number, by the way, is an **exponent**.

The same thing works for very small numbers. The Planck constant is about 0.0000000000000007 eV.s, which is pretty hard to work with, but we can write it as:

7×10^{-16} eV.s

You'll sometimes see these written like this:

10^{14} written as 10E14, or

10^{-32} written as 10E-32

The main reason being that it's not always possible to use little numbers (like on a calculator screen, for example) so people just put in an E (for 'exponent') instead.

Multiplying and dividing with exponents

When you multiply powers of 10, you just add the zeros, like this:

10 x 100 = 1000

(one zero times two zeros = three zeros)

This works no matter how many zeros you have:

1000 x 1000 = 1,000,000

And when you divide, you can just take away zeros:

10,000 / 10 = 1000

or

1,000,000 / 1000 = 1000

When you're working with exponents, you just add or subtract, like this:

10^6 x 10^3 = 10^9

(that's 1,000,000 x 1000 = 1,000,000,000: 6 + 9 = 12)

It works just as well with very large and very small numbers:

10^7 x 10^{-3} = 10^4

(that's 10,000,000 times 0.001 = 10,000: 7 - 3 = 4)

Algebra with exponents

Okay, let's put all of this together.

Example

I've just measured the concentration of sugar in my swimming pool and it's 3×10^{-2} grams per cubic meter. I often accidentally drop mugs of tea into my swimming pool without drinking them (actually: I just do it for fun). If my pool cleaner can pick up one empty mug from the bottom of the pool each day, it's 10 days since the water was last changed and there are still four mugs at the bottom of my pool, how much sugar do I put in each mug? My pool is 4 m x 2 m x 20 m and is completely full.

Answer

0.34 grams

Explanation

Let V be the volume of my pool

Let S be the amount of sugar in a mug

Let M be the number of mugs dropped in

Let P be the number of mugs picked out by the pool cleaner

Let R be the number of mugs left at the end of the period

Let D be the number of days in the period

Let C be the concentration of sugar in the pool

$C = 3 \times 10^{-2}$

$C = M \times S / V$

$P = D \times 1$

R = M - P

D = 10

R = 4

V = 4 x 2 x 20

What is M?

Let's start by simplifying some stuff:

V = 4 x 2 x 20

V = 160

C = M x S / V

V = 160

so

C = M x S / 160

Now we're trying to find S to let's get it on its own on one side of the equation. Multiply both sides by 160:

C x 160 = M x S / ~~160 x 160~~

C x 160 = M x S

Now we'll divide both sides by M:

C x 160 / M = M x S / M

C x 160 / M = S ~~x M / M~~

C x 160 / M = S

or in other words:

S = C x 160 / M

Now let's substitute for C:

S = C x 160 / M

$C = 3 \times 10^{-2}$

$S = 3 \times 10^{-2} \times 160 / M$

$S = 4.8 / M$

Can we substitute for M to get rid of it as well? We've got:

$R = M - P$

Let's add P to both sides:

$R + P = M \cancel{- P + P}$

$R + P = M$

$M = R + P$

And we know that $R = 4$, so:

$M = 4 + P$

P is 10 (since $P = 1 \times D$ and $D = 10$) so we can substitute that as well to get:

$M = 4 + 10$

$M = 14$

Okay now we can go back to our equation for S:

$S = 4.8 / M$

$M = 14$

$S = 4.8 / 14$

$S = 0.34$

So ... 0.34 grams of sugar in each mug.

Roots

The opposite of squaring something is taking the square root of it.

So $3^2 = 9$

And $\sqrt{9} = 3$

In other words, doing $\sqrt{}$ (called "taking the square root") is the opposite of squaring.

In fact, if you do one after the other, you end up with the number you first started with, like this:

$\sqrt{(A^2)} = A$

And also:

$(\sqrt{A})^2 = A$

Remember how we could 'balance' an equation by adding, subtracting, multiplying or dividing both sides by the same thing? Well you can take the square or the square root of both sides as well, all without breaking anything.

Here's an example:

$A = 3.14 \times r^2$

$A = 49$

What is r?

Let's substitute for A:

$49 = 3.14 \times r^2$

Now we'll divide both sides by 3.14:

$15.6 = r^2$

Now we'll take the square root of both sides:

$\sqrt{15.6} = \sqrt{r^2}$

The right-hand side is just:

$\sqrt{15.6} = r$

And we can use a calculator to find the left-hand side:

$3.95 = r$

In fact, **you can do anything to an equation, as long as you do it to both sides, and you won't break anything**.

Fractions

Yes, I know, fractions are scary. They were scary in primary school, and they still scare people.

It's all very well to look at this:

2 / 3

But when you start looking at things like this:

$$\frac{4}{5} + \frac{3}{4}$$

... they're not so friendly any more.

As with equations, there are some simple rules with fractions that you can use to tame them and find the solution. They are:

you can add or subtract fractions if they're the same underneath (ie if the stuff under the line is the same on both of them)

you can multiply or divide the top and bottom of a fraction without breaking it as long as you do the same to both bits

Let's look at rule 1. Here are some fractions that you can either add, or not, depending on whether they're the same underneath:

$$\frac{7}{4} + \frac{1}{4} = \frac{8}{4}$$

$$\frac{7}{4} - \frac{1}{4} = \frac{6}{4}$$

$$\frac{7}{7} + \frac{1}{4} = (not\ the\ same!)$$

$$\frac{7}{156.94} + \frac{1}{156.94} = \frac{8}{156.94}$$

$$\frac{7}{A} + \frac{1}{A} = \frac{8}{A}$$

$$\frac{7}{A} + \frac{1}{B} = (not\ the\ same!)$$

$$\frac{7}{(A+B)} + \frac{1}{(A+B)} = \frac{8}{(A+B)}$$

$$\frac{C}{(A+B)} + \frac{D}{(A+B)} = \frac{(C+D)}{(A+B)}$$

$$\frac{C}{(A+B)} - \frac{D}{(A+B)} = \frac{(C-D)}{(A+B)}$$

Now let's look at rule 2.

$$\frac{7}{4} + \frac{1}{8}$$

We can't do this addition because they're not the same underneath. But what if we multiply the top and the bottom of the left-hand bit by 2:

$$\frac{7 \times 2}{4 \times 2} + \frac{1}{8}$$

$$\frac{14}{8} + \frac{1}{8}$$

Now they're the same, and we can add them:

$$\frac{14}{8} + \frac{1}{8} = \frac{15}{8}$$

We can do the same with variables:

$$\frac{7}{4} + \frac{1}{4 \times A}$$

Just multiply the top and bottom of the left-hand bit by A, like this:

$$\frac{7 \times A}{4 \times A} + \frac{1}{4 \times A}$$

Then we can do the addition:

$$\frac{7 \times A}{4 \times A} + \frac{1}{4 \times A} = \frac{(7 \times A) + 1}{4 \times A}$$

Multiplication signs

Although we were all brought up using the 'x' symbol to mean 'times', mathematicians hate it. The reason is very simple: there's a tradition of using the letter x as a variable name, and it's too easy to confuse x with x.

So in equations, mathematicians often replace the x multiplication sign with something else. Sometimes they replace it with a dot '.' (which to my mind is just as confusing). Sometimes they leave it out altogether, which even *more* confusing!

Here's what it looks like in each of those forms:

4 x A

4.A

4A

They mean *exactly* the same thing. I guess the last version has the advantage of needing less ink, which has to be a good thing.

Let's look at that fraction example again, just so you can get the feel of the new (missing) multiplication sign:

$$\frac{7}{4} + \frac{1}{4A}$$

$$\frac{7A}{4A} + \frac{1}{4A}$$

$$\frac{7A}{4A} + \frac{1}{4A} = \frac{7A + 1}{4A}$$

Notice that we can drop the brackets around things like 7A in "7A + 1" because there's less possibility of confusion than with "7 x A + 1".

Algebra with fractions

Now it's your turn. Here's an example for you to try, that involves fractions. I've numbered the equations, too, just to make it easier to follow the explanation:

Example

Equation 1: $A = \frac{D}{B+C} + 6$

Equation 2: $A = B + C$

Equation 3: $A = \frac{D}{2}$

What is A?

Answer

A = 8

Explanation

From equation 3, we can find D:

$A = \dfrac{D}{2}$

Multiply both sides by 2:

$2A = D$

Now substitute that into equation 1:

$A = \dfrac{\boldsymbol{D}}{B+C} + 6$

$A = \dfrac{\boldsymbol{2A}}{B+C} + 6$

Now substitute from equation 2:

$A = \dfrac{2A}{\boldsymbol{B+C}} + 6$

$A = \dfrac{2A}{\boldsymbol{A}} + 6$

Divide the bottom and top of the fraction by A:

$A = \dfrac{2A}{A} + 6$

$A = \dfrac{2}{1} + 6$

So the answer is 8.

Final example

Okay, one final example that will (I hope) bring together most of what you've learned so far. The Doppler effect is what makes an ambulance siren go first up in tone, then down in tone as it rushes past. It's caused by the fact that the ambulance is travelling first towards you, then away from you, fast enough for the sound waves to be compressed or extended as you hear them.

The Doppler equation is:

$$f = (\frac{c + r}{c + s})f_0$$

Where:

f is the frequency that you actually hear

c is the speed of sound in air

r is the speed you are travelling at

s is the speed of the ambulance (the 'source' of the sound)

f_0 is the frequency of the siren on the ambulance

If:

- the speed of sound in air ('c') is 340 m/s
- you are standing still
- the siren is actually at 350 Hz, but
- you hear the siren as 300 Hz

Then how fast was the ambulance travelling?

Answer: 60 m/s

What's next?

If you'd like to see what other books I've written, you can do that here:

```
amazon.com/author/philcohen
```

Some of what I've written has been fiction (but not this book, honest!) but you'll also find my books on accounting and selling, and one on how to escape from a job you hate.

Printed in Great Britain
by Amazon